WHY SHOULD I EAT THIS?

PROTEIN

TABLE OF CONTENTS

By Lindsay Snea

A Crabtree Seedlings Book

Crabtree Publishing

crabtreebooks.com

School-to-Home Support for Caregivers and Teachers

This book helps children grow by letting them practice reading. Here are a few guiding questions to help the reader with building his or her comprehension skills. Possible answers appear here in red.

Before Reading:

- What do I think this book is about?
    - *I think this book is about food.*
    - *I think this book is about protein.*
- What do I want to learn about protein?
    - *I want to learn what foods have protein.*
    - *I want to learn why it is important to eat foods with protein.*

During Reading:

- I wonder why...
    - *I wonder why some people don't eat meat.*
      I wonder how tofu is made from soybeans.
- What have I learned so far?
    - *I have learned that protein is good for our muscles and bones.*
    - *I have learned that we can get protein from foods that come from both animals and plants.*

After Reading:

- What details did I learn about protein?
    - *I have learned that eating scrambled eggs gives our bodies protein.*
    - *I have learned that we should eat foods with protein every day.*
- Read the book again and look for the vocabulary words.
    - *I see the word **legumes** on page 10, and the word **develop** on page 12. The other glossary words are found on pages 22 and 23.*

PROTEIN

Protein is a **nutrient** found in food. Proteins build your muscles and bones. They help you to have healthy blood and skin.

Protein also helps keep your **immune system** strong to protect you from getting sick.

You can get protein by eating foods that come from animals and plants.

Protein from animals can be found in meat, fish, and eggs, as well as in dairy foods such as milk and cheese.

Protein from plants comes from foods such as **legumes**, nuts, and seeds.

Protein helps your body grow and **develop**. It is important to make sure you eat foods that have protein every day.

DID YOU KNOW?
We need different amounts of protein at different ages.

You can get protein through snacks such as yogurt, cheese, peanut butter, sunflower seeds, or a glass of milk.

Meals with protein could include scrambled eggs for breakfast, a ham and cheese sandwich for lunch, and salmon for dinner.

Vegetarians are people who do not eat meat. They get their protein by eating foods from plants, such as lentils, tofu, and nuts.

DID YOU KNOW?

Tofu is a food made from a kind of bean called a soybean.

Eat protein every day for a stronger, healthier you.

Glossary

develop (dih-VEL-uhp): To get bigger or stronger

immune system (ih-MYOON SIS-tuhm): The system in your body that protects you from getting sick

legume (LEG-yoom): A type of food, such as peas or beans, that grows inside a pod

nutrient (NOO-tree-uhnt): A substance that your body needs to live and grow

vegetarian (vej-i-TAIR-ee-uhn): A person who does not eat meat

Index

About the Author

Lindsay Snea lives in the mountains of Fernie, British Columbia, Canada, with her husband and four children. Lindsay is an elementary school teacher, and a lifelong learner, who is passionate about cooking, food, fitness, being outdoors, and health.

Written by: Lindsay Snea
Designed by: Margaret Salter
Proofreader: Melissa Boyce
Photographs: All images from Shutterstock
Production manager: Candice Campbell

Crabtree Publishing

crabtreebooks.com 800-387-7650

Printed in the U.S.A./CP112025

Published in Canada
Crabtree Publishing
616 Welland Avenue
St. Catharines, Ontario
L2M 5V6

Published in the United States
Crabtree Publishing
347 Fifth Avenue
Suite 1402-145
New York, New York, 10016

Library and Archives Canada Cataloguing in Publication
Available at Library and Archives Canada

Library of Congress Cataloging-in-Publication Data
Available at the Library of Congress

Hardcover: 978-1-0398-8117-4
Paperback: 978-1-0398-8478-6
Ebook (pdf): 978-1-0398-8237-9
Epub: 978-1-0398-8357-4
Read-Along: 978-1-0398-8562-2
Audio: 978-1-0398-8520-2